Train like a Scientist, Think like an Entrepreneur

How to leverage your research training to excel in the start-up world

By

Stacy Chin, Ph.D.

Disclaimer

The information provided in this book is not a substitute for professional, legal, or financial advice. The author makes no legal claims, and the material is not intended to replace the services of an attorney or the requirements of governmental agencies.

ABOUT THE AUTHOR

Stacy Chin is a champion and advocate for researchers and innovators whose passion is to turn their idea into a viable business. Dr. Chin, co-founder and CEO of Hydroglyde Coatings and a distinguished polymer chemist, shares her personal and professional journey from the laboratory to the board room in her book, *Train Like a Scientist, Think Like an Entrepreneur.* An NIH grant reviewer and mentor to first-time founders, she has set the bar high, made the mistakes, experienced the highs and lows, struggled to go on, and been rewarded with success. Her message—and personal discovery—is that what scientists have been trained to do at the bench, they are uniquely equipped to do in the marketplace: Bring an idea to life. The key is to learn the language of entrepreneurship and the mindset that governs it.

A passion for research and an entrepreneurial spirit drive Dr. Chin's innovative endeavors. Her start-up, HydroGlyde Coatings was born out of a mission to deliver biomedical solutions to advance contraceptive products and global health. This led to the creation of the first-ever self-lubricating condom coating technology that received funding support from the Bill and Melinda Gates Foundation and National Institute of Health.

Dr. Chin's interdisciplinary research background spans across polymers, biomaterials, biochemistry, structural analysis, and coatings. She secured over $1M of grant funding and is also the co-inventor of over a dozen granted and pending national and global patents. She has published and co-authored manuscripts in JACS, ACS Macro Letters, and Chemical Science. During her doctorate training, she was recognized for her research efforts through several awards and fellowships. She earned her chemistry Ph.D. at Boston University in the Grinstaff Lab and her BA from the College of the Holy Cross.

Dr. Chin consults start-ups on their business, non-dilutive funding, and research strategies. Dr. Chin holds leadership positions in R&D, product development, and business development, as well as manages interdisciplinary teams and collaborations with foreign partners. She has worked with founders, professors, medical doctors, and collaborators from top research institutions and hospitals, including MIT, Mass General Hospital, Columbia University, University of Colorado, UCLA, Brown University, Boston University, New York University, University of Texas, UCSD, and more.

Train like a Scientist, Think like an Entrepreneur

TABLE OF CONTENTS

INTRODUCTION

Five of us waited in the reception area. I was number 3. Each of us had 10 minutes to prove our idea was worth $100,000. Gazing out the massive windows onto the Boston Harbor skyline from the 20th floor, I was tenser than I had been during my doctorate qualifying orals, if that was possible.

Breathe, remember your lines, stay calm—and don't talk too fast.

The double doors clicked opened and the sound made me jump. The two remaining candidates looked at me then a voice called out my name. *Stacy Chin.* My heart skipped a beat. Ten seconds later I stood in front of 12 men in high-back executive chairs neatly positioned behind a crafted oak table, all dressed in expensive tailored suits with polished designer belts and shoes.

Don't screw this up.

It was not my first rodeo. I had delivered dozens of scientific presentations, published multiple articles in scientific journals, submitted patent applications based on my scientific findings, and was acknowledged for my research efforts through several awards. So when I was finally standing in front of reputable investors with a newly-minted Ph.D. and a prototype proposal backed by the Bill and Melinda Gates Foundation and the National Institute of Health (NIH), I expected to be well-received. It was only later that I came to understand that in the eyes of start-up savvy investors, a doctorate in polymer chemistry was equal to wearing the Scarlet Letter, proclaiming: *She is a scientist, not a business leader.*

Nevertheless, each one dutifully scribbled on their yellow pads during my seven-minute presentation filled with non-stop charts, graphs, and numbers. My hard-core technical slides were solid for a doctoral defense but had no structure or storyline to spark the slightest interest to these deep-pocketed investors. Worse, the financials were completely off (never again will I use the "hockey stick" projection!). When I finished my slides, I said thank you and politely asked for questions.

> *Do you have a pro forma and financial projections?*
> *What is the estimated market capture in the next 3 to 5 years?*
> *Do you have a regulatory strategy?*
> *Have you considered whether you are B2B or B2C company?*
> *What is your market and advertising strategy?*
> *Are you licensing your technology or launching a brand?*

This was the moment of truth. As a scientist, my world was all about forming structured hypotheses and testing them. Every day I approached the bench using highly controlled experiments to answer questions quantitatively using statistical analyses. At that time, my "market research" was reviewing the scientific literature and reading papers in my field. But here I was suddenly facing a wilderness with new rules and laws governing it—some I would learn over the next several years were directly opposed to the scientific training I had dedicated nearly a decade to achieve and perfect. I felt like a first year-graduate student again—a "G1" as we were called then.

Charting new horizons

Like me, you are a scientist and you understand. You've spent years in a laboratory working 10-12 hour days, 6-7 days a week, honing your skills, generating and analyzing data, accumulating knowledge with the aim of becoming an expert in a narrow area of research. You've fiddled around with massive analytical equipment, from mass spectroscopy to nuclear magnetic resonance. You've poured color-less liquids tirelessly into glass containers from massive liter-sized solutions all the way into microliter plates. You've demonstrated amazing dedication and adjusted to having little or no downtime in your life because getting your Ph.D. was your world.

Since then, maybe you've developed a prototype that could disrupt the medical device industry or invented a new small molecule with promising therapeutic effects. Or you're nearing the end of your postdoc fellowship and have a big decision to make: *Should I re-apply, get a traditional academic job, or pursue my ground-breaking start-up idea?*

Or perhaps you've already put in some time in an industry but now you're more than ready to move on and create something of your own. Either way, you're confident in your capacity to succeed. You've leveraged your decade of research training to troubleshoot (that time when no one had a clue) why the freshly synthesized organic 100 mg sample had turned bright blue when it should have been white. You've operated large chromatography instrumentations to figure out why the red alarms were going off and the internal microfluidic pressures were out of whack.

Bottom-line: You know you need to acquire serious business acumen. Scientists without a base knowledge of business will not be successful as founders or CEOs. I'm not just talking about the typical roadmap for putting together a start-up: prototype development, market analysis, ownership, or even fundraising. Resources to do that are readily available: excellent books, articles, YouTube videos, and courses abound. What I'm talking about is something more essential, elusive, and rarely addressed—a business mindset.

My road to entrepreneurship

My involvement in commercial development began when I started a research project backed by the Bill and Melinda Gates Foundation. We were a strong technical team of scientists and engineers. But none of us had a business education. What we did have was resilience in the repeated face of failure, a scientist's key strength. The disadvantage—and a serious one—is the difficulty most scientists have in seeing the big picture and knowing how to spearhead the business—which is, by necessity, the main focus of a start-up.

My focus on the technical details is the reason my pitch fell flat and died an instant death before I was even out of the room. I failed to present the big picture, to tell the story of the problem being solved or the need being met. I stayed in my scientific comfort zone: data, data, data and results, results, results. That's what I knew, what I'd spent years training to do.

Over the next three years, I shifted gears and dedicated myself to learning everything I could about start-up entrepreneurship and business. I read hundreds of books on marketing, product development, pitching, regulatory policy, and financing. I found supportive mentors, inspiriting advisors, and, finally, excited investors who encouraged my transition from scientist to CEO. I attended events with experienced CEOs and business leaders and learned to network—to talk and behave as a business professional. Along the way, I earned a Harvard business certificate and began to develop leadership and communication skills—something graduate school does not teach—and I continue to learn.

Eventually with trial and error, we did raise over $1.4 million in government funding from the National Institute of Health and other institutions to develop a prototype and prepare to go to market. Yet, after that traumatizing pitch, I did consider playing down my scientific training although I would never go as far as some people who deliberately leave their Ph.D. title off their presentations. In 2018, our start-up and I were awarded the Entrepreneur of the Year Award from the Worcester Chamber of Commerce. We secured patents, created a prototype in eighteen months, secured critical partnerships, and prepared for a human market trial study.

Since then, I've had the opportunity to share my experience at competitions and local incubator panels. The most frequent question I get asked is: *What was the most challenging part of your transition from scientist to CEO?*

Before answering, I always take a moment to proudly say that I regard myself as a scientist. That it was my scientific training, in fact, that served as a foundation for my business development. Then I point to the key differentiator for success: learning how to leverage my scientific training mindset into a founder's mindset.

As research experts, we are rigorously trained to answers questions, to be factual, to showcase our knowledge. Our ability to do so is a sign of confidence, ability, and experience. During my qualifying oral exam, for instance, when I didn't know the answer to a professor's question, he laughed out loud at me—*ugh, how humiliating*. So when you come to be a founder or CEO, you believe you should know the answer to every question or problem. The most challenging part of your transition from scientist to CEO will be your ability to openly admit what you don't know—and this, for scientists, goes against the rules.

The best people and mentors I have worked with are fully confident in what they do know and are also quite confident in what they do not know. To put it simply: In science, you desire to answer every question; in business, you don't. So if you're ready to travel this road, I offer to be your guide. As a scientist, I understand the challenges and I've walked some of the way. What I'm going to share is my experience of how I learned to develop a business mindset to complement, not replace, my scientific training.

The key areas covered in this book:

- How to adapt your mindset from researcher to entrepreneur

- How to fight and keep going through times of doubt
- The top 3 pitfalls you have to navigate as a scientist
- Why seeing the big picture is the main focus of a start-up
- Why soft skills are the most important skills in business leadership
- How to effectively communicate your commercial idea
- How to approach the challenge of taking your scientific research into a start-up
- Why the first members of your A-team will determine your start-up's future
- How to decide which fundraising strategy to use, and
- Why self-care and knowing your limitations are the key markers for founder success.

You're ahead of the game

Contrary to what you may have heard, your academic training as an independent scientific researcher can be your key to success. Objectivity, methodology, creativity, and persistence in the face of failure: I discovered that these major advantages allowed me to function successfully in the lab *and* the boardroom.

Chapter 1
GETTING THE START-UP STARTED

Whenever you see a successful business, someone once made a courageous decision.
-Peter Drucker

I never had the goal of becoming a CEO of a start-up. I thought I wanted to be a scientist. In 2013, I was selected as the lead chemist of a 4-person technical team on a project sponsored by the Bill and Melinda Gates Foundation whose well-known mission is to solve the world's most pressing global health problems. Five years later I became the Founder and CEO of HydroGlyde Coatings. The purpose of this research project was to develop a new condom coating technology—a device that had not altered in 70 years—in order to reduce the global spread of STI's and unplanned pregnancies. Like any scientist who dreams of—and then gets an opportunity—to make a difference, to relieve suffering and prevent disease, I was thrilled and could barely contain my anticipation.

Having met the first couple of requirements of a start-up—fill a commercial real-world need, put together a team and obtain seed funding—we were ready to begin prototyping in the lab. In this phase, few would argue that the scientist has a clear advantage. Scientists are innately curious people and deep analytical thinkers. We enjoy questioning the world around us. To satisfy our craving for knowledge, we seek answers through hypothesis-driven experiments. The challenge of finding the truth is exciting and enjoyable to us. There is nothing more fulfilling when experiments actually work, outcomes finally are clear, results can be statistically interpreted, and conclusions can be made. This process may seem rational and straightforward to the non-scientist. Yet in reality, getting experiments to go even (moderately) smoothly at the bench can take weeks, months, or even years.

The Prototype Journey

First, we had to determine our goal. What exactly *were* we trying to do?

Here's what we knew: Condoms are accessible and cheap. They are the only product that can prevent disease while serving as a contraceptive. Yet people don't use them because of low user satisfaction. *So, how do we make them better*? Retail research was our first step. We purchased and examined every type of condom we could get our hands on and then read up on every article and textbook about coatings and relevant material science content. We soon recognized that condoms are not lubricated sufficiently which leads to discomfort, resulting in inconsistent use or no use at all. *Bingo*. Our aim became clear: to create a thin, self-lubricating coating technology to encourage people to use condoms more consistently and correctly.

We were ready to go into the lab and develop a prototype. We dove in and tried countless coating formulations. Weeks passed, months went by, and nothing worked. Then a year and still nothing. It felt like we were climbing Mt. Everest, every day we hit a new snowstorm, steering us off track from the summit of innovation which began to seem unattainable. We were weary and had to fight hard against deepening discouragement. Yet we knew that making innovation happen takes patience and requires time to work through the technical kinks and the endless trouble-shooting.

Did I titrate my buffer with the wrong reagents? Are my CHO cell line contaminated with mycoplasma? Did I use too much cyclohexanes instead of ethyl acetate in my chromatography column?

Day-in and day-out we retraced our steps. We reviewed, questioned, and critiqued every procedure to confirm what went right, what could have gone wrong, and how to move forward and try again. And we kept going.

Two years in, we felt the pressures from our team, advisors, and friends who told us to move on. *Move on to what?* I argued. We had nowhere to go. Did they mean it was time to give up? At this point, we had registered our company and landed some governmental funding, but we were still a technical team without any expert business guidance. (We had tried out a couple of consultants, only to be overcharged and given no clear direction forward.) It seemed we were at a crossroads and maybe the advisors were right. *Was it feasible to keep going?*

The judgment call was a difficult one. So far we had created upward of one thousand formulations and although it was unlikely, and perhaps unwise to attempt any more—especially from the outside, looking in—the team collectively decided to pursue one more prototyping round. The next Saturday afternoon, mentally prepared for another failed experiment, I mixed a couple of solutions together, put the coating formulation on the condom, added water, and touched the surface.

It was . . . slippery. *Whoa, I think we did it!*

I rushed to tell the rest of the team who were equally shocked and then euphoric upon verifying it—we finally had a formulation that worked!

New game, new ground

With the initial prototype breakthrough, we were over the moon, eager to bring an actual product to market. We approached several manufacturers excitedly and quickly signed an agreement with one we believed was a good fit (although we didn't fully understand all of the terms in the agreement at that time). Securing a reputable manufacturing partner for large-scale prototyping was a huge step. *Now we had to figure out how to translate our scrappy bench-top application process into a real-world product.*

The manufacturing plant was a scientist's playpen, filled with countless ways to test and improve our formulation. It was a fascinating challenge that fed our curious technical brains.

If we change the rate of the dipping process, does it increase the thickness of our coating? If we adjust the production speed moderately, can we still achieve consistency between each unit?

But in reality, we weren't at the bench anymore. We were working with unfamiliar processes and restrictions governing them—as well as burgeoning costs. As stated in the contract, each prototyping cycle would cost two months and thousands of dollars—not to mention another round of questioning from impatient investors and our other partners waiting in the wings. With time and money closing in, decisions became more difficult.

As lead chemist, it was a constant internal battle. *Do we revise the formulation again for the 124th time? Do we have enough data to move forward? What happens down the road if we don't pursue this experiment? Do we have enough time and funding to try these 3 more experiments without delaying our milestones and timeline?*

At a critical point, things started to fall apart. The technical team was in disagreement, funding was running out. Then our manufacturing partnership fell through. We were in crisis mode with no idea what to do next. Why were we still pushing forward then? We were afraid not to. With limited time and funding, it's often necessary to execute without knowing all the details or facts. Plus we thought naively: We were scientists, couldn't we figure it out? We failed to understand the vulnerable position we were in.

But the real reason came down to this: we didn't know what we didn't know. We needed help.

LESSONS LEARNED: Know when to take risks

Successful business owners or investors understand when to take a risk and *how to balance that decision with the potential reward*. For start-ups, this is a common struggle, especially in the early make-or-break days. Consider the challenges we faced in both situations—developing the prototype and partnering up with a manufacturer.

- **In the initial prototype development**, the choice to follow our collective urge, to keep going, was based on our years at the bench formulating, testing, analyzing. This was territory we commanded with confidence. We had learned to face failure and regroup: that was our strength. So while this choice might not be considered a calculated risk (especially by non-scientists), it was a reasonable one.

- **Partnering with a manufacturer**, on the other hand, was an unwise risk because it was outside the experience and expertise of the team. We had no training in developing processes that would make it possible to translate the coating to thousands of devices at a time under highly regulated testing parameters using ISO standards. The misstep here was our unexamined assumptions that we could somehow figure it out because we were scientists—and our serious lack of guidance.

Next Up: THE BIG PICTURE

How to get the kind of help you need when you reach your limitations.

Chapter 2
THE BIG PICTURE

Starting a business and building a product are not for the faint of heart. You have to learn to not let little disappointments get you down and to stay focused on the big picture.
-Gillian Tans, Chairwoman, booking.com

It was our fourth meeting and I was early. I wanted to settle in before I popped the question. Laptops splayed open everywhere, the café was full of mid-morning murmurs and the familiar clang of drinks being made. I usually sipped my latte slowly but I had drunk it all down already. I got up to get another and the sun broke through the morning grayness, casting a shard of light across my notebook. A good omen, I hoped. I checked the time: 10:12 am. Eighteen minutes before show time. He was always prompt.

Start-ups falter for many reasons: you miss a critical deadline, run out of money, lose a key founding member, your expected financing falls through, or your first customer (or in our case, manufacturing partner) backs out. It's a paralyzing moment for founders when you recognize that you and your team have reached your limitations. For us, it was the wake-up call that something needed to change. We needed help, we needed business expertise, we needed to take the *righ*t next steps. What we really needed was a Board Chairman. Someone who had "walked the walk" before, someone who could help us clarify our vision, and someone who could help us chart our direction and strategy. My gut told me this was the only way forward. I knew we needed to build out our team, but I wasn't sure where to begin as my network only consisted of chemists and engineers.

Where were we going to find talented people with business expertise?

Thankfully, I discovered that Boston has a rich start-up network. (If your area doesn't, thanks to the rise of entrepreneurship, there are many online resources.) Now I just had to find a way to tap into it. I began by attending events sponsored by accelerator groups, incubators, investors, and other entrepreneurs. I met investors who had helped dozens of start-ups raise millions of dollars and grew companies over 10-, 25-, and even 100-fold. I met business executives who led start-ups towards multiple exits and IPOs. I met marketing and sales professionals who launched products across the US and globally. At each event, I made it a goal to meet at least five new people and set meetings with three of them. Although I wasn't too sure who or what I was trying to find, I knew I was at least heading in the right direction.

For months, each week filled up with meetings. I sipped on lattes with business executives and consultants and enjoyed trying different lunches with marketing professionals and legal advisors. I presented what I thought were the most promising features of our start-up. We had a viable solution to address a serious global health problem with backing by the Gates Foundation and NIH. We had a new innovation to disrupt a stagnant industry with growing market potential.

As I realized through my conversations, many on the other side of the table offered a broader perspective. While recognizing the value of the problem we were solving, they challenged me to think on a macro-scale, with questions like:

How much is the market opportunity?
How many existing products are made globally?
Who are the competitors and what is their market capture?
What is the costs analysis of the product supply chain?
When do you anticipate revenues?

Eventually, as these questions repeated themselves, I saw the pattern and addressed them upfront as much as I could. Some meetings were dead ends and others led me to my next step, with variations on a similar theme:

"*You should speak to this person.*"
"*I may be able to introduce you to someone who can help.*"
"*Have you reached out to that person?*"

The weeks rolled by and the coffee and lunch meetings led to exponentially more coffee and lunch meetings to the point I lost my love for coffee—but I got better at honing my elevator pitch and describing our vision. I introduced promising candidates to our team, and in a few instances started discussions about bringing them onboard, only to have them choose other opportunities. Other team members brought in candidates as well, but it was never the right fit. One month, then six months passed and we felt beaten down and discouraged. After endless meetings, I was tired of the same familiar words and attitudes that didn't stir me. Some were reserved, arrogant, others chummy and endlessly probing—but it all felt the same. No go. No door opened. And how would I know if it had? I had no basis, no experience to judge. But I was looking for something and I hadn't yet found it.

Then finally I met an experienced investor at a university pitch event. From the beginning, he seemed genuinely interested in our venture and excited about our idea. Like everyone, his resume displayed an outstanding career in business and sales. It was clear he had the expertise and reputable network we desperately needed to develop our team. But this particular individual was different—he offered more. He was encouraging, present, creative, optimistic, and enthusiastic and, unlike most other candidates, he acknowledged and appreciated the value of my technical background, regarding my Ph.D. as a strength rather than a weakness.

He introduced me to a new way of thinking, what he called the Big Picture—with questions like: *What is your Plan B if your Plan A doesn't work out?* He challenged me to re-shape my perspective and mindset to think of the emerging, daunting start-up challenges we faced as exciting opportunities for exploration and growth. At the end of our third coffee meeting, I felt more than hopeful. Here was someone I was comfortable with, someone I could openly discuss our start-up challenges with. Someone who was willing to spend the time and energy to mentor me and coach me. Someone who shared a similar vision and was eager about the opportunity.

I was ready and waiting.

"Hey, Stacy." I heard his voice behind me and I turned.

"Good morning."

"What are you having this morning?"

I always had the same thing, a large latte with almond milk, and he knew it. It was his way of saying hello and showing his interest.

We got seated and I blurted it out.

"I want you on our team. How do I bring you onboard?"

He looked at me, smiled, and said the words I will never forget, "You know, Stacy, I'm going to take a chance on you."

These are words I will forever remember. Onboarding him as my board chairman was one of the best decisions of my start-up career. Not only has he helped to transform and shape the start-up to where it is today, but he has invested in me professionally and supported me to become a start-up executive. Not many board chairmen will do this. I am immensely grateful.

On your start-up journey, you will encounter excellent mentors and experts who are willing and interested in helping you out. They will share their knowledge and connect you to their network which will help you to build your start-up. Keep a look-out for another type of mentor that I call the Sponsor. Sponsors do more than advise. They are willing to roll up their sleeves and contribute—and join you to make your start-up become the best it can be. Sponsors walk besides you, teach you, coach you. They can help you develop your ability to think strategically and see the bigger picture. When you meet a mentor or advisor that you sense could become your sponsor, put in the time to nurture the relationship. This kind of support is invaluable and rare—and for founders on their first journey out, a tremendous boost.

The Big Picture

As scientists, you are trained to focus on carefully crafting each individual puzzle piece. Entrepreneurs, on the other hand, focus on relating each individual puzzle piece to fit into each other—in order to generate the big picture. For the scientist training to be an entrepreneur, it can be daunting to transition from being an expert in one field to someone who is a complete novice or beginner in a completely separate field. That's okay—and it's necessary. The key step is to recognize is that you're on a whole new playing ground and while knowing that your scientific skill sets are valuable and essential—you must also be willing to grow beyond them.

LESSONS LEARNED: Asking new questions

You can't do it all. The sooner you realize that and identify the right resources to address your gaps, the faster progress your start-up can make. To grow, you have to let go. Here are some proven guidelines to begin that journey.

- **Get out of the lab.** Your idea and your company will not get off the ground without top-notch business guidance. With strong strategic direction early on, you can avoid wasting time and money in floundering decisions at your highest risk point.

- **Be a beginner.** Setting aside your ego is necessary to be a beginner. It allows you to open your mind fully to what you don't know so you can learn that much faster to see, and make decisions based on the Big Picture.
- **Don't settle**. Your business relationships are as important as your business idea, maybe more so. Seek out a network of advisors and mentors. Be thorough in your search until you find the right person to guide you. Capture a sponsor if you can.
- **Choose qualities along with talent.** The critical factor in building a solid business team is choosing the *RIGHT* team members who are not only trustworthy, motivated, energetic and passionate, but also who share your start-up vision. These characteristics are as important as experience, achievements, and business expertise.

Next Up: FOUNDER'S BLUES

How to realign your technical team to do business.

Chapter 3
FOUNDER'S BLUES

Being a founder is great for personal growth, because if you don't grow, your business dies.
-Sahil Lavingia, Gumroad Founder & CEO

The transition from a technical team to a fully functioning business organization is like a Grand Prix pit stop: you need to change tires and refuel fast and efficiently. And if you're competing in an endurance race, your success might even mandate a driver change as well.

One of the most challenging, and least-acknowledged problems that scientific founders aren't prepared to deal with (and we were no different) is team realignment: the creation of a business hierarchy. This is a critical stage at which many start-ups fail. Even with an experienced chairman or wise investor guidance, interpersonal conflicts and struggles cannot be avoided. It's doubly difficult because the team has been together for a while and gone through a lot. They are bonded—and that bond is being altered and maybe even threatened. Naturally, you don't want to upset or offend the team member whose hard work and support have gotten you through hard times—and yet, as leader, you're responsible for making decisions that will move the company forward. People will be reassigned, asked to leave, and move on—and loyalties will shift.

The process of decision making also changes. Some members may not understand or agree with the rationale behind new decisions in which they no longer participate or play a diminished role. *EGO*. We've all got one, and when roles change, people react unexpectedly. Hurt feelings, anger, and resentments are likely to occur, even when the process is well-managed. Another stress factor often overlooked is your technical team's lack of business experience—in the middle of all these changes, they must also adapt and function in a new environment.

In the early start-up stage, usually the CEO and the chairman are responsible for managing this often chaotic and emotionally difficult process. As CEO, and with our chairman's guidance, I had to steer the process and deal with people's resistance, as well as my own inner conflicts. As a scientific founder going through this daunting obstacle for the first-time, I was grateful to lean on his support throughout our restructuring process. Without our chairman's guidance and leadership, our team would not have transitioned as smoothly as it did. I've also gained a better sense of awareness of my own weaknesses and strengths, allowing me to grow faster to adapt to this new role.

Our story

When our chairman came onboard, the first order of business was to persuade the founding team to take a good hard look at our roles and each other's involvement in the company. We had to ask—and answer—the difficult questions:

- Which technical founders will move forward?
- Who needs to step back?
- How will roles and responsibilities shift?

- How will all these changes impact everyone's share of company ownership (the capitalization table)?

Re-negotiating the cap table

For our start-up to have a chance for success, we realized that every member had to be committed full-time. Some team members had commitments to stay in academia, a typical situation for start-ups founded from academic research. One founder had committed to MIT for his post-doctoral training; another founder was a full-time professor. So after some discussion, these two agreed to step down. The remaining founder assumed a new role leading our engineering efforts.

The final—and emotionally-charged--issue was adjusting the cap table to reflect a fair balance of ownership and voting rights. We had to reflect and discuss what future work would be done by whom and how much we anticipate committing our efforts to those future responsibilities. On the one hand, it was important to acknowledge the departing members for all their work which had brought us to this point—and yet, on the other, it was necessary to consider what percentage was appropriate since they were leaving the company.

Re-negotiating the role and company ownership between the professor and me was the most challenging. This is a common problem in academia where a professor (or as I've witnessed, multiple professors) often takes a huge chunk of a company in the early stages. Investors do not want to see professors on cap tables who are not active members of the company. As the leading founder and CEO, I needed majority ownership whereas the professor needed to accept a significant reduction of ownership. It was the only way to move forward and we both knew it.

Having been a Ph.D. student in his lab, I had no idea how to even begin this conversation. As my advisor for the past five years, I looked up to him as a mentor, a teacher, an advocate. Under his guidance, I had become an independent researcher. We had a great relationship and I didn't want to jeopardize that in any way. Without him, the start-up would not have ever happened!

Our chairman understood my concerns and we agreed I might not be the appropriate person to address this situation with him. Quite often, many start-ups that find themselves in this position seek outside 3rd party counsel, such as lawyers or investors, to help mediate this conflict. After dedicating blood, sweat, and tears, founders develop personal connections to their start-up, as well as each other, making it incredibly challenging to objectively tackle this without having emotions and ego get in the way. Because of his experience and negotiating skills in dealing with a situation like this, our chairman spoke to the professor one-on-one. Thankfully, the professor understood and agreed to step back for the successful future of the company, giving us his best wishes and support.

Do you want to be a CEO?

There is considerable debate among investors about whether the lead founder is a good choice for the role of CEO. Technical founders are passionate and very focused: two characteristics that are regarded as a big positive by some investors or a huge negative by others. However given the lack of business training, technical founders may have a difficult time outlining the business strategy and constructing the high-level direction for the company. The bottom line: If you are so far into the weeds that you can't pull yourself away from the fascinating details to lift your head and take in the full impact of the business horizon, your company won't be able to grow beyond its fishbowl existence.

Many technical founders seek out the primary role of entrepreneur; some don't. It's not the only option. You may want to assume technical leadership or license your prototype and move on to something else. But before you back away from the CEO role because you lack experience or doubt your abilities—both my initial weaknesses—take a close look at what you bring to the party.

Start by realizing that as a scientist you have a singular advantage: you already have some of the attributes and skillsets essential to be a successful CEO. Drive, analytical skills, innovation and persistence in the face of failure (resilience) are the most valued and needed. Unlike a business-trained CEO, you also possess deep technical knowledge and invaluable skills developed from your years of study, research and experimentation. However, you will have to learn to put your natural craving for knowledge aside in favor of the "Big Picture," which can be challenging for some with a scientific mindset. You will face the necessity to make decisions without adequate time to measure all the results. Approximation may be all you have in some situations and you have to learn to be okay with that. As I've learned from my mentor, sometimes you just have to lock the engineer (or your inner scientist) in the closet.

Other key skills you will need to develop are the ability to motivate and manage others, communication and negotiation, and knowledge of business structure and its functions (legal, financial, human resources, sales and marketing). Most agree that the hardest thing about being a CEO is managing people. Some would even argue that the most challenging situations involve your trusted, fellow team members! You may know how to manage a technical team in a lab or research setting but dealing with people with conflicting goals and functions under pressure in a business landscape is very different. Books and courses will get your through a lot but it's experience that will ultimately teach you.

Know yourself

Another way to view the role of the CEO is through temperament. Knowledge and skills can be acquired—temperament is more difficult to alter. It requires constant self-examination and self-awareness (more on that later).

There was never an "Aha!" moment when I knew being CEO was right for me or that it was even what I wanted. But as time went on, it became clear that I was willing to take risks and be vulnerable, to put myself on the front line without knowing what I was supposed to do or how to do it. It was my desire to make a go of our company, my confidence in our idea and its potential to help others that moved me forward. I didn't realize it at the time but *I had the temperament.* And what I didn't have I was ready to develop, which included finding my own leadership style.

From the outset, I discovered that being a CEO is a humbling position. With support and practice, I learned to reflect on my decisions, to admit my errors, and to ask the hard questions.

LESSONS LEARNED: Restructure your start-up and know yourself

Transforming a team of scientists into a functioning business is complex, difficult, and chaotic. Here are some examples of leadership and negotiation skills I've learned that made our transition possible.

- **Do not leave the room**.
  When founders are engaged in discussions on how to re-organize the team and make decisions regarding roles and company shares, do not leave the room until everyone is in agreement *and* everyone has a smile on their face. You do not want to burn any bridges.
- **Re-negotiating cap tables.**
  To get founders to agree to cut back their shares, you might need to ask this question: "Do you want 10% of a company that is worth 0 dollars? Or do you want 5% of a company that will be worth $100M?"
- **Seek outside perspectives.**
  Scientists can sometimes be overconfident—or blindsided—in their analysis of a situation. I've found it essential to pay attention to other perspectives. For example, I listened when investors warned me

that raising funds was a no-go when professors held large shareholder percentages.

- **Know yourself.**
  Being CEO isn't for everyone. Take the time for self-reflection to see if you have the temperament and the desire to be on the front line and take responsibility for the direction of the company. Can you also be open about what you need to learn and do you have the willingness (and humility) to admit your mistakes?

Next Up: THE BACKWARD MINDSET OF COMMERCIALIZATION
Navigating the territory between prototype and product.

Chapter 4
THE BACKWARD MINDSET OF COMMERCIALIZATION

Experience indicates that even the most promising ideas face daunting
commercialization challenges and a nearly impassable Valley of Death
between promising concept and viable product.
-David M. Anderson, Agile Product Development for Mass Customization

I recently met a doctor who had an idea for a wash basin that would make treating wounds easier and more efficient. He said he had 400 prototypes in his garage but he needed more money to refine his prototypes and bring down manufacturing costs. When I asked him if he had validated the product with other doctors, he said no because he wanted to get the cost down first. The idea of getting feedback from medical professionals who would use the basin did not seem significant to him; in fact, it was not part of his process at all. Yet, this step is actually the most important of all.

Scientists are not known for their commercial mindset. Asking questions and following the path of inquiry—*what happens if we try this?*—is what we do best and love to do. Most of us have spent the years of our first career incarnation in an academic laboratory environment formulating and testing a series of hypotheses. So when you, as a scientist, don the entrepreneurial cap and step out of the lab with a prototype in hand with the goal of commercializing your "product," you do the scientific thing: you ask questions. The next logical step? You find answers.

You learn as much as possible about the commercialization process as you can, from prototype to FDA or other regulatory approvals and everything in between. Hiring consultants, taking business courses, going to conferences, reading books and articles, applying for grants, seeking investors, and looking for manufacturing and distribution partners are all viable options.

Getting knowledge and information is important and necessary. However, lack of knowledge is not one of the top reasons most start-ups, especially from academia, fail to go beyond the prototype stage. The more common reason is the inability to change our mindset. Mindset defines a certain more or less fixed way of thinking. It's how we approach problems. A scientific mindset tends to be methodological and logical, following procedures in a certain, prescribed manner.

If you are thinking like an entrepreneur, you don't start with a question or a methodology: you begin with the end goal—the customer. Then you work backward to achieve it within a set of milestones and deadlines. This ideology is the complete opposite of your training as a scientist. So how do you shift your mindset?

The Lean Start-up Method

The standard way to bring a new product to market for decades has been to develop a business plan, get funding, hire the full team, including the CEO. This approach was lengthy (months into years) and ate up a lot of resources before the proof of concept could be reached.

A company would develop and launch a product that made zero ripples in the market or was outright rejected by the customer—and was then forced to go back to the drawing board or abandon the project.

That method was challenged by Steven Blank, veteran entrepreneur and current champion of the scientist as CEO, who encouraged founders to quickly validate the customer and the market before going through the lengthy process of full development. This "lean" approach accelerated proof-of-concept by allowing multiple innovations through the pipeline for testing and consideration by the potential market. The goal was (and is) to achieve a Minimal Viable Product (MVP) as soon as possible *before* manufacturing or other expensive processes are undertaken.

Thus the Lean Start-up Method was born.

Consider our doctor's wash basin scenario. If a group of medical workers tests out the basin for treating wounds with real patients, the inventor is no longer working blind; he can make additional iterations that fit the needs and function of the customer he's designing for. Steven Blank sums up this process as "... [the] lean-start-up method says that you don't come here to execute a business plan. You come here because you have a set of untested hypotheses."

Adopted by the National Science Foundation, the method is now well-known and widely practiced and applauded by grantmakers, VCs and other investors as a sound basis for successful commercialization. The reason for its success is that it encourages experimentation and accelerates development while minimizing resources. This is where the ideas of "fail fast" and "pivoting" originated.

This method is based on the following principle: *How quickly you can test your product and validate the market will determine your success.*

Making the shift

In the academic research environment, the scientist is the master of carefully crafting experiments with full control over variables and methods. If you are lucky, you have the luxury of time to spend weeks, months, and (often) years to test a hypothesis. You have the freedom to allow your intellectual curiosity to branch out into other investigations to expand your understanding. You have the confidence that you're making decisions based on "complete"—near perfect as possible—data or results. In a lab experiment, for example, we must have a 95% significant change or result before moving forward! This is the beauty and value of science and research.

Yet, when prototyping new technologies in the start-up world, these luxuries do not exist. With dwindling time and money against your odds, you have only a few tries to get the experiment right. This is where the logic and efficacy of the lean approach, once understood, makes sense to most scientists, allowing them to make that necessary "mindset" shift. By changing the way you look at your data or the latest iteration of your prototype, you go from: "Is this as perfect as possible?" to "Is this good enough to move forward?"

Our story

As scientists, we are used to surveying the scientific literature before initiating a new research project: what's been done, what were the outcomes, what are the current opinions or theories of leaders in the field, and what is the next stage opportunity for development. From the entrepreneurial perspective, this "opportunity" must be defined in terms of the problem you are proposing to solve in the real world and whether the market even wants your solution.

At Hydroglyde, that problem is a global one: helping to lower the one million sexually-transmitted diseases contracted every day by developing a self-lubricating condom. In early stage development, we proved our first hypothesis: that it was possible to create a new coating designed to make the condom more lubricious without the need for silicone lubricant. Prototype in hand, we faced the task of commercializing for the consumer marketplace:

- Test and transfer the technology for human use (testing/manufacturing) and
- Validate the market (human trials).

A word about markets—make sure you know the space you are in from the beginning. This will save you so much time, money, and effort if this is done properly from the get-go. When scientists are asked *'What is your market?"* they can turn to the available literature. For example, if your "product" is a medical device, you get the latest copy of "The Global Medical Devices Market Report" and read it thoroughly, priding yourself on the fact that you now know that the market, your market—is driven by the increase in the incidences of infectious and chronic diseases—*and* is expected to grow at a CAGR of 6.1% from 2021 and reach $603.5 **billion** by 2023. That's the market, right? Well, not quite.

There is a basic division within the marketplace: the business-to-consumer market (B2C) and the business-to-business market (B2B). The requirements for commercialization are vastly different between them. Both can be quite difficult to enter, bringing their own unique set of challenges. For example, a lot of start-ups jump in right away to get FDA approval for their working prototype. Sometimes that makes sense. Other times it might not. Either way, you must lay out the best path forward to get market validation testing done first, especially if your product is in the consumer market space.

Manufacturing Partner

Our first step toward commercializing our prototype was to get a top-notch manufacturer. Finding the right one is critical in the product world. These people are your partners, your right hand, and your path into the market. Without them, you have nothing—literally.

In our arena, there are hundreds of manufacturers in our industry across the globe. Picking the right one for us was like finding a needle in a haystack. Beyond learning about standards and regulations (a lot of them), supply chains and distribution, we had to put equal value on developing trust and rapport. It didn't make it easier that our checklist for a suitable partner was quite long:

- Technical expertise in our product category

- Committed to making a high-quality product for consumers
- Adherence to strict medical device regulatory standards across different countries
- Shared our vision and values
- Believed in the potential and enthusiasm about what we are doing
- Willing to roll up their sleeves and work with us to get to the end goal
- Trustworthy and transparent
- Shared the same urgency as we do to get our product to market as soon as possible.

Don't rush through this step—make sure you get the guidance to understand the kind of support you need and are thorough in your due diligence. The first time around we failed on both points. You don't want to put yourself through that kind of disappointment and wasted time.

At first, I didn't know where to begin. So I created a list and started cold-calling to see what kind of reception I got on the other end. *Were people interested, willing to talk to me?* From there, I narrowed down the list to three or four and had the next level discussions. We toured multiple facilities, had more phone calls. Finally, we got on the plane and went to Asia. We went through more rounds of negotiation and contracts. This took lots of time, lots of effort, lots of travel, and lots of searching. However our efforts paid off. In the end, we onboarded a trustworthy manufacturing partner that saw our vision and wanted to come along on the journey.

Human Trials

Our next phase was to plan and prepare for human trials—the all-important phase of validating the market. In our first round, the plan was to introduce five couples to a new technology in a product that hadn't seen any changes in 70 years! We were excited and naturally nervous about the outcome. Founders want—and may even expect—the customer to love their product, so a certain amount of detachment is required, which is admittedly difficult when you've spent years developing your "baby." It's also important to be prepared for the reality that start-up products (especially consumer products) require a lot of tweaks and iterations to become "marketable."

To manage this process with human studies, I recommend separating the founders from the participants. To ensure objectivity, we onboarded a highly-experienced marketing firm to administer and set up the trials. Here are the questions we needed answered:

- Is our prototype better than what's currently being used?
- Will people like it—and want to use it?
- How many iterations will be needed – do we want to set a limit?
- What are our costs and timeline?

Unlike performing an experiment in a lab, in product testing you have little or no control over many of the variables. No tweaking of well-defined experimental variables like formulation concentrations, temperature, or switching out reagents. You're dealing with humans—and the kind of variables scientists would never find on the bench top: feelings, emotions, opinions, past experiences. Yet these all-important variables are what make up the consumer experience and ultimately will determine your success or failure.

A big challenge we faced was that condoms in general have a negative connotation. Knowing that, we realized that when couples saw our product, they might have mixed reactions. Getting background of their previous experience and history with competitive products, and learning some of their pain points helped us to put into perspective how they might judge our prototype.

As expected, our first round produced a range of different types of feedback. We were of course delighted with the positive feedback. But just like the many "failures" we've all experienced at the bench, it's the negative feedback that's the most important since that's what determines your next improvement.

LESSONS LEARNED: Beyond your fairy tale script

Taking the leap into commercialization is exciting *and* full of rude awakenings, as every founder discovers. It's easy to get seduced and fall in love with our idea, our prototype, our vision. Passion is a good thing—but take the feedback, adjust your mindset. Get ahead (and stay in) the game by adopting an entrepreneurial mindset as soon as possible.

- **The scientific approach (most of the time) does not apply to product development.** Spending time to "perfect" a prototype *before* getting marketing validation in most cases is a guarantee for failure. The sooner you have your MVP in hand and customer validation, the better chances you have for successful commercialization.

- **Don't be afraid to pick up the phone and get on the plane**. You need to research and verify for yourself what you're getting into when bringing on manufacturing or other critical third-party support. Having the right people on-board to support your journey is critical.

Next Up: **THE MISSING LINK: SOFT SKILLS**

Playing hard ball with soft skills.

Chapter 5
THE MISSING LINK: SOFT SKILLS

The soft skills are the hard skills. People who master the critical leadership skills today are anything but touchy-feely — they're direct, they're clear, they're compassionate, they're no-nonsense. But they're not soft.

-Amy Edmondson, Professor of Leadership and Management, Harvard Business School and author, The Fearless Organization

A study by the Massachusetts Biotechnology Foundation (MassBioEd) recognized that leadership ability represents a big gap in scientists' research training: *"...scientists with little background in managing people are often hired or promoted into leadership positions and find themselves dealing with peers and superiors with equally limited managerial experience."*

Why do scientists score poorly when it comes to leadership ability and good management skills?

As scientists, our training goal, day after day and years after years, has been to master the hard, technical skills—understanding complex problems, forming strong hypotheses, perfecting experimental protocols, analyzing large amounts of data. Given the vast differences in scientific training from university to university, and even from lab to lab, most scientists have not been trained to communicate their science effectively. Yet these are the most critical skills needed to articulate and communicate complex technical data. Ironically, and greatly to our disadvantage, only a lucky few get the opportunity to fully develop their presentation and writing skills.

There are reasons for this gap.

During the course of completing a research masters or doctorate, most scientists generally get limited practice and few opportunities to present their work, both to peers and the scientific community at large. "Group meetings" are the most common form of presentation practice where scientists in training share their research developments to their lab mates and advisors. These 15-to-30-minute presentations typically happen two or three times a year. Likewise, most graduate students struggle to get the proper practice and training to master technical writing for grant proposals, peer-reviewed publications, and even their thesis before heading off into the real world. Technical writing is very challenging as it takes practice and time to learn to articulate and streamline dense and complex ideas on paper.

Thanks to my liberal arts undergraduate education, I got a head start in developing my presentation and writing skills. Because of this training, my Ph.D. advisor worked closely with me to help write academic grants and fellowships to the National Science Foundation and the National Institute of Health. By the end of my doctorate, my professor and I had written over a dozen proposals. At the time I had no idea how critical these writing and communications skills would be in launching my start-up. Without these skills I would never have secured over $1M in grant funding that had put us on the road to becoming a real company.

A problematic mindset

The training gap for most scientists, however, represents more than a lack of communication skills. It reflects a mindset: a mode we learn early on, a style of communicating in which we are trained to defend our position and focus on providing data and facts to back it up. As a result, scientists are generally preoccupied with proving their point and can find it *very* challenging to be receptive to the differing views and ideas of others.

This mindset is also reinforced by the fact that scientists train mostly inside an isolated and competitive environment, often working alone or having little contact with others. Collaboration and teamwork, although encouraged, are sometimes avoided due to the fear that colleagues or lab mates might scoop your research findings and claim them as their own.

It's not surprising then that a scientist, suddenly promoted to a leadership and management position, find themselves out of their element, in deep water where they can barely do the dog paddle much less swim laps. Compounding the situation, many scientists ignore this deficit, unaware of the detrimental effect it can have on their career as they transition to managerial positions in their professional life.

If you do have the rare opportunity to work openly and collaboratively with other scientists and engineers—sharing ideas, listening to other perspectives and ways of approaching problems, and being challenged about your own methods—you are fortunate and more prepared than most to succeed in managerial or leadership positions.

The value of soft skills

Seth Godin, noted author and branding specialist, recently published an article entitled "Let's stop calling them soft skills." This garnered wide attention from many business and start-up consultants such as Kate Leto and Alison Entsminger, both of whom lead prominent training and hiring programs in emotional intelligence, resilience and self-awareness. The article argues that since 'soft skills' are the key to success in leadership and perhaps even in life, why are we calling them soft especially when they are so hard to acquire.

Valuing soft skills doesn't mean you reject or put your "hard" skills on a lower plane. In fact, the deep competencies acquired in research—independent and logical thinking, analytical and number skills—give you a key advantage that non-technical leaders do not have. Whether working with the people on your team, an important vendor, a potential customer or an investor, your knowledge and understanding of a technical issue or problem may be a tipping point in making a decision or coming to a consensus. However, for your perspective to have an impact and be well-received, it must be *coupled with your ability to communicate it effectively.*

The productivity that comes from a happy marriage of soft and hard skills cannot be underestimated. When I began serving as a proposal reviewer for a governmental agency that funds science start-ups, I was bowled over by the exceptional communications skills and collaborative mindset of these mostly technical leaders and scientists. Before coming together to discuss a proposal and reach a consensus about recommending it, each of us reviews the proposal independently and develops our own opinion as to whether the project is fundable. Different opinions are shared and heard by the others; no one tries to grab the stage and convince the others of the rightness of their view. Instead, we are able to have very engaging and open-minded technical conversations, sharing our viewpoints with mutual respect. Sometimes, an argument is so convincing that after giving an open hearing to another perspective, reviewers change their original evaluation.

The collaborative mindset demonstrated by these reviewers and the communication skills that support it, are the same soft skills you will need to move your start-up forward. Your ability to listen and consider different views and perspectives with an open mind is what will allow you to manage conflict and work productively with your team.

The bottom line: Realize from the get-go that if you are serious about turning your vision and your company into reality, soft skills are as important, if not more important, than your technical knowledge. This is a big mindset shift for scientists. *Huge*. The next step, also huge, is to realize that, like your scientific skills, soft skills must be cultivated; they will not happen automatically. And without them, your start-up is likely to join the long list of post-mortem statistics.

How to develop soft skills

To acquire soft skills, you will need self-awareness, a conscious effort and a commitment to your own growth. It's a challenging journey, especially for scientists, one that forces us to be honest and vulnerable with ourselves, as well as those we work with.

So, how do you go about developing soft skills? There are lots of ways but the main strategy is to get outside your routine and expand your social interactions by putting yourself in different environments and situations you're not accustomed to or even comfortable with. Put on your scientific hat and observe how different people respond and communicate with each other. Use this same lens to observe how you respond and communicate with others as well.

Begin with self-examination and ask yourself:

How do I engage and interact with others?

What are my strengths and my weaknesses in my communication style?

Am I a good listener?

How do I communicate with others and response to those with different viewpoints?

What situations cause me to react defensively?

Some may find a coach or mentor to be very helpful in this self-examination process. Consider getting involved in your community by volunteering for a leadership position. I joined a couple of non-profit boards that encourage women and minorities in STEM which was very helpful. I learned by watching how others expressed themselves and engaged with the other members, and most importantly, how they handled disagreements while respecting conflicting views.

Ask someone you admire to be a mentor. Read books. Take professional and personal development courses. Hang out with colleagues and peers who have these skills. Put in the hard work. Going on this professional development journey is one of the biggest investments you will make in yourself, your start-up and your future.

LESSONS LEARNED: How to speak both languages

Soft skills form the core of leadership ability. *Period.* Whether you are hiring or managing your team, negotiating with an investor or important supplier, your ability to share your vision, communicate a goal, and most importantly, problem-solve with your team, will depend on the strength of your soft skills.

- **Put soft skills at the top of your list.** Strong communication and negotiation skills and the ability to listen and support your team members are not "extras" or "nice-to-haves." These critical skills will make or break your professional or entrepreneurial life.
- **Hear out different perspectives.** Acknowledge a person's view and express your own. If you label the views of others as wrong or incorrect, you're going to make it more difficult for your own views to be heard and acknowledged.
- **View yourself as an experiment.** Find a way to get excited and curious. Join an organization in your community where you can observe and practice soft skills—and give yourself permission to ask questions.

Next Up: BUILDING YOUR A-TEAM

Your start-up is only as good as your team.

Chapter 6
BUILDING YOUR A-TEAM

Choose the first 10 employees very carefully, as they determine the next 100.
-Yevgeniy Brikman, Gruntwork

Every start-up founder faces an overload of critical decisions early in the life of their company. None are as important as selecting the first few people you choose to work with: Your A-team. These are the talented and passionate individuals who will work tirelessly toward the common goal of translating your dream, your idea or product, into reality.

These first individuals will form the foundation of your company. These are people whose depth of experience will help you craft a business strategy that will forge your start-up's future. Most importantly, these are people who, having gone down this path before, know what it takes to succeed.

So, how does a founder start the hunt for these key players?

The big divide

Finding the right people for an early stage start-up is very different from finding people for an industry job. Aside from the fact that you are not a big pharma company with an HR department to put out the job description, gather in all the resumes and set up the interview—the people who walk through those traditional doors may not be the people you're looking for. Why not?

Many first-time candidates are excited about the idea of working for a start-up. Based on how the media portrays it, these candidates mostly view start-ups as exciting environments where they can use their talent to work on innovations to tackle some of the world's biggest problems. Although this is true, what they may not be aware of (and what the media doesn't capture) are the long nights and weekends, the management challenges, the fear as to whether you can expect next month's paycheck to put a roof over your head, and the stress that comes from the pressure to meet tight and changing deadlines. Early stage start-ups can be emotionally, mentally, and physically draining. And this impacts everyone involved differently. During the high seasons, you will observe the best of the best of people. In the low seasons, you will encounter some of the worse sides in people as well.

It takes a particular kind of person who can stomach the up's and down's and the demands, the sudden pivots of the start-up world. Someone with a strong curiosity and drive who can tolerate failure and re-starts: whether it's the 10th or 100th time around and still has the resilience, optimism, and energy to get up and try it again. These qualities are not on a resume; they are part of a person's temperament.

On the bright side, many newly-minted Ph.D.'s actually fit this description already. They are used to the long hours, stress, deadlines, and repeated failures. They've worked on new innovations on the bench top as well. Overall, they have the talent and temperament that a start-up needs to succeed. However, there's a dark side to the start-up culture they may not be aware of, another aspect the media doesn't portray.

Like many bright and eager researchers coming out of graduate school, a colleague in my Ph.D. lab thought the start-up environment would be ideal for her. She loved research and enjoyed innovation and the freedom to pursue scientific studies. She believed it would be a place where she could make a difference in people's lives. However, when she joined a 5-person start-up as one of their lead biomedical engineers in her first job, she was taken aback—and totally unprepared for the dynamics and culture that often exist in a start-up space. She expected that everyone would have an equal say only to find a division between the business and technical teams. Neither team was interested in her input nor was she allowed the creative freedom to explore and direct her research. Instead she was required to work on the experiments already planned. Tight deadlines had to be met which changed constantly. When technical discussions became heated, people were not kind.

The long hours she had worked as a graduate student continued and she was often required to work on the weekends as well. It didn't take long for her (and the team) to realize that this particular start-up was not the right fit for her. After resigning, she found a job at a large pharma company that fit her and she loves it there.

It's important to note that not all start-ups have similar expectations or dynamics that she experienced. Many leaders (and their teams) take care to create a mindful and respectful environment that supports teamwork and thus creates a strong foundation that will move the start-up forward. As the leader, it is your job to retain talent so make sure you provide those around you with a supportive environment. And for those looking to enter start-ups, it is very important that you can tolerate that start-up's particular culture and determine whether that is the right environment for you to grow and thrive.

A custom fit

Unlike traditional hiring methods, putting together your A-team is a very personalized process. For example, your first board member, your technical lead, your business strategist: these roles must be a custom fit, nothing off-the-shelf will do. The countless considerations and pressures can be overwhelming so your first priority is to build a team that will complement you: your leadership style, your values, your abilities and skills, how you work day-to-day, how you make decisions.

Now's the time to reflect on the kind of people that will work best with your personality and give you the support you need. Like many founders, I faced the narrowing window of time, money and opportunity, thus the need to onboard a diverse team quickly. Under such pressure, it's easy to make mistakes—to choose people who are not compatible with you, personally or professionally, or the rest of the team. That's why it's essential to take the time to know candidates beyond a couple rounds of interviews and reference calls, as these are the people you will work closely with, who will be involved in major decisions affecting your future and the future of the company.

Seek people who fit the start-up profile and then some: who thrive in a fast-paced environment, who are able to work independently and also in groups, can handle last-minute weekend meetings and "crazy" changes in plans and schedules, who love challenge and don't need a lot of hand-holding. In short, people with the kind of foresight and strategic savvy who will turn your single start-up engine into a charging and unstoppable locomotive.

My goal was to find people with abilities and experience that filled in my gaps and weaknesses, people who were passionate and aligned with my values, smarter than me, with whom I could also risk vulnerability and admit mistakes, trusting they would stick around and move forward with me. I needed people I trusted and could depend on if I were to call at 10 pm on a Saturday night because the problem or opportunity at hand could not wait till morning.

Next: Consider the key functions and roles you need. Naturally, that will depend on the stage of development you're in. If you're working on a prototype, you probably don't need marketing or sales expertise just yet. If you're ready for commercialization, however, you probably need support for: IP, manufacturing, finance/legal, marketing and product development.

For every position, you must be able to answer the question: Do I understand exactly what is needed? The skill sets, experience, talent? Is knowledge of my industry important? Identify both the role and the goal. For example, you may know the expertise and experience you want your technical candidate to have, but what is the specific goal you want this person to achieve or contribute to in their first six months or year? Then list the qualities and soft skills you think are essential: negotiation, writing and presentation skills, teamwork. If you're not sure, seek out someone who has gone down the road before you, an experienced person you trust.

The value of a network

Traditional hiring methods—such as posting job descriptions and collecting resumes—most of the time do not work well for early stage start-ups because the caliber of people you're looking for goes far beyond even the most qualified employees. Even if you have the resources to hire a reputable, expensive recruiting firm or a HR professional, it may not be the best place to look. In fact, relying on typical "hiring" avenues can actually restrict your hiring pool and you might miss out on some incredible talent.

In my experience, word of mouth is often the most productive path to building your A-team, especially the first one or two members (who can then lead you to the third, fourth, or even tenth person that will round out your team). The strength of a network lies in the people who have first-hand knowledge of each other's personality and abilities, strengths and weaknesses. What I realized early on in my search for a board chairman was that most talented people are sought and known within the network. They get so many opportunities to join boards, be advisors, or accept executive positions that they don't need to put out a resume! These are the people you want.

Our chairman's extensive network was critical in building the team that would support our company towards commercialization. After working together to figure out the experience and skillsets we needed, he introduced me to potential candidates who could bring outstanding value. I was amazed at the people I met. Not only were their resumes incredibly impressive but they also mirrored the high energy, compassion, dedication, teamwork, and strong work ethic that we both shared.

Still it can be hard to find the perfect person.

For instance, a key position we needed as we approached human trials was a marketing advisor. Because our unique product falls under two categories—medical device and consumer product—we needed someone with experience on both sides, no easy feat to find. After days of Linked-In searching and pinging our network, we were frustrated by the lack of suitable candidates. We interviewed a lot of people with impressive resumes, but the right person did not emerge. Finally, one of our board members told us about a marketing professional who had a marketing and consumer product background and had worked in the field of medical devices. As a former executive of P&G, she brought experience working with consumers and had mentored lots of start-ups. She ended up being the perfect fit.

It took several months to go from four technical founders to a highly diverse team of 10. But one by one, we onboarded new advisors and members with all the right skill sets and qualities we needed. Business strategy—check. Negotiation expertise—check. Legal—check. Medical devices—check. Financial strategy—check. When my team and company finally felt right, I felt unstoppable.

LESSONS LEARNED: Hire people with the right stuff

It's impossible to overestimate the importance of having the right team in place to bring your idea—and your company—into the kind of alignment that is unstoppable.

- **Hire slow.** I've made poor hiring decisions, as well as good ones—and the experience really hammered home how critical it is to onboard the right people. Poor hires that do not fit the company culture, are not aligned with your values, who do not understand how to spend their time and resources properly will drain and emotionally exhaust you and cause ill-will and resentments. Worse, such distractions can dangerously stall your progress by keeping you from the work at hand: moving your start-up forward.

- **Custom fit.** The key factor in building your A-team is to tailor it to your needs, abilities, and temperament as the founder. This requires a lot of self-awareness about how you work with others, what your communications style is, and how you make decisions. My strength is that I know myself well. I know what ticks me off, I know how I work with different types of people and I know how I respond to them.

- **Recruitment shortcut.** When it's time to kick your company into major gear, consider other avenues other than posting ads on Linked in, Hacker News, Product Hunt, Mashable, or other start-up job boards. Instead, work with an advisor/consultant who has a vital and experienced network in your space.

Next Up: BRIDGING THE GAP

Got your team assembled? Now you have to lead them forward.

Chapter 7
BRIDGING THE GAP

The question I ask myself like almost every day is, 'Am I doing the most important thing I could be doing?'... Unless I feel like I'm working on the most important problem that I can help with, then I'm not going to feel good about how I'm spending my time.
– Mark Zuckerberg, Founder and CEO, Facebook

Recently I consulted for a diagnostic company caught up in a conflict that resulted in a stalemate of an important project. The start-up needed to prepare a white paper about a COVID-19 test they were developing that uses a proprietary protein to detect parts of the virus. If all went well, this could lead to a partnership with a well-known Fortune 500 company. The business team was naturally excited about this opportunity and wanted to highlight and "beef up" what they thought were the research accomplishments and then some. The technical team resisted, as they wanted to only present the facts. After some rounds of discussion with both sides, it was clear that neither side understood the language of the other. The business team did not understand the difference between proteins and nucleic acids. The technical team did not understand why their technical results were re-written to sound like marketing copy.

Eventually, after hours of heated discussion and "translation," the sales team was forced to accept the fact that technical results they were pushing to present could not be claimed since these experiments had not actually happened yet! Lots of time and effort went into these conversations. As a result, the white paper was put on hold and the Fortune 500 company partnership fell through. Had the communication been clear from the beginning, the conflict could have been avoided and all that energy could have been used to push the start-up forward.

Bridging the technical and business gap

By necessity and design, your A-team consists of diverse professionals with different perspectives, whose passionate beliefs in those perspectives are backed up by experience and long practice. (That's why you hired them!) Yet, you and the team are working together toward the same goal—a working prototype, financial support, a proven market, a successful launch—and that's the priority you need to stay focused on, even as you consider the various views and ideas for accomplishing your goal.

Having a strong technical team is critical in a start-up. Having a strong business team is also critical, especially as you expand from a purely technical focus to become a full-fledged business. However, the perspective, strategy, communication, and language for both sides are very different, as the above story demonstrates. What the technical team may find important may not be important at all to the business gurus; the milestones outlined by the business team may not make sense to the technical team.

Handling disagreements and conflict among team members while directing them toward the higher goal beyond their personal or professional preferences is a big order for any leader. This can be especially challenging for the technical founder whose "first" language is science. Due to my lack of business acumen, especially in my early days as scientist-turned-entrepreneur, I was intent upon learning everything about a business problem or issue before I made a decision. I soon realized it was not only impractical—there wasn't enough time. More importantly, as CEO, this was not my role. I needed to rely on my A-team to help me with these key decisions.

So how does a technical founder bridge the gap between these two different sides that requires both types of mindsets, different skill sets, different ways of looking at problems, different people, and different ways of communicating?

Speaking (and listening) in both languages

Bridging the gap between the technical and business team may seem like a daunting job. However, time and time again, people have said (and I have seen it demonstrated) that it is easier to teach a scientist business than a businessperson science. Effective communication is the solution to this challenge and it needs to start with you. You, as the technical founder/CEO, are the key to bridging this gap. To do it well you must learn to speak, and respect the vastly different languages of the business and technical worlds and eventually become fluent in both (or close to it).

At the top of the list is your ability to listen carefully to both sides as dispassionately as you can—listening both for the gaps and the misunderstandings that arise out of them, as well as for the potential solutions they offer to the problem at hand. Doing this requires you to *interact closely with your team members' strengths and limitations.* Many scientists find this a difficult and even a distasteful territory.

This is where your soft skills will count; without them you will flounder.

For example, your finance officer may want to hold back buying some piece of equipment that your lead technician believes (passionately) the company can't do without. How do you, as the company leader, bring them to a consensus? Listen, then ask questions and encourage other team members to do the same. As a scientist would do, gather all the knowledge (the data) from both parties and identify how you can bridge these gaps. Search out and expose assumptions that may cause misunderstandings. It may turn out that the equipment won't actually be needed until the next developmental phase is completed—or it might be that the phase itself can't be completed without it. Search out alternatives—there are often alternatives that suggest themselves but you must look for them.

The qualities for effective communication that are most needed are an open mind, your willingness and ability to hear all views and to consider them carefully without favoritism or pre-conceived assumptions. When you give each team member a full and thoughtful hearing, you are modeling the behavior you want them to have toward each other—behavior which supports collaboration and teamwork.

At the practical level, your role initially may be to act as translator (and sometimes ambassador) while not getting caught in the middle as referee. Encourage the technical people to see the bigger picture that the business people have laid out. Likewise, you can help the business people understand why a technical process may take longer than expected because more time is needed to explore X, Y, and

Z. Once all views are on the table and well-sorted, the optimal direction may become clear. But if it's not, your role is to lead a productive discussion that rises above debate and focuses on how the differing views can lead in the direction of a solution. Depending on the issue, it may be best to have all hands-on deck to brainstorm next steps. If it's a strictly technical issue, you may want to select the relevant team members.

Your goal is to avoid being caught in the net of conflict or confusion, except, in reality of course, you will get caught. You will be flummoxed and not understand why your technical lead keeps pushing to run another experiment even though last week everyone agreed (including him/her) not to. Or, why your business manager wants to begin negotiations with a manufacturer or demo a potential customer even though the prototype is not ready to show. These instances can be quite challenging to mitigate, especially as the founder. However, never rush into a resolution.

When you and your team feel an urgency to act, you may have to slow yourself (and them) down to make sure that everyone is on the same page and understands the issues at hand. Resist the urge to rush headlong because backtracking from a mistake is something you want to avoid; retracing your steps is often demoralizing and expensive, so try not to do it.

Since it is the founder's job to unify the team so the start-up can move forward, there are times you will find yourself in the challenging position of having to convince one side or another to get on the same page when they both disagree. Another important part of unifying the team is to keep everyone motivated by assuring them that their participation and feedback are valued.

So how can a technical founder help to marry the two worlds of science and business together?

Leverage your strengths—know your weaknesses

Traditionally as scientists, we are valued and known for our independent thinking and scientific contributions. Business leaders are sought after and rewarded for their skill in building and leading a strong team to the endgame of success. In today's start-up world, the challenge of the scientist-entrepreneur encompasses both: to bring the rigors of science and the opportunities of business together. What is the path? And how will you traverse it successfully?

Start by appreciating the value of your scientific training.

The scientist excels at discovery. She is an expert at holding an open space for the unknown. The scientist weighs, measures, tests, observes and does it all again to discover—and generate—new results. Throughout this process, she must keep an open, observing mind ready to encounter the unexpected and the unanticipated. The underlying drive that informs this exploration sparks a continuous stream of questioning and experimentation.

When you, as a scientist, take on the role of founder or CEO, you don't leave the science behind. Quite the opposite: your scientific training, used wisely, is a critical asset that will propel your start-up forward. Much of what the scientist learns in their day-to-day training actually resembles very closely what goes on in their start-ups—but be forewarned that the strengths and skills acquired at the bench top can also become liabilities.

Here's how to deal with them.

1—The scientist possesses the discipline of focus and concentration.

You've spent months and even years focusing deeply on solving problems. Your ability to keep your eye on the ball will be a key advantage as CEO in leading your team forward.

RISK: Getting lost in the weeds.

WHAT TO DO: Remind yourself to come back to the surface and reconnect with the bigger picture. Is the problem you're working on getting you closer to the end goal?

2—The scientist knows how to handle failure.

How many times have you planned an experiment and executed it only to find that 99% of the time, it veers off in unexpected direction, forcing you to regroup, create a new plan and run it again?

RISK: Don't know when to stop.

WHAT TO DO: Figure out how to pivot. It is good to try something again to see if it works but don't get sucked down a rabbit hole when there may be another solution. Take a step back. Check in with yourself and your team.

3—The scientist is an innovator.

You have spent your scientific training doing ground-breaking research. You have the adventurous curiosity to keep asking the hard questions and the courage to challenge assumptions.

RISK: Failure to recognize the limits of innovation to solve every problem.

WHAT TO DO: Innovation is exciting and motivating. But it doesn't always solve the most important problem or fill a real need. At the end of the day, it's all about the customer. Do you have a buyer that wants the innovation? Make sure the innovation solves the problem for the end user.

4—The scientist makes decisions based on evidence and data.
You advance in your goal and avoid mistakes by going in the direction the evidence leads.

RISK: Ignore insights or dismiss the value of what cannot be measured or evaluated scientifically.

WHAT TO DO: Realize that decisions are not always based on scientific evidence or results. In fact, sometimes they have nothing to do with science at all. It's another dimension called business.

LESSONS LEARNED: Take the high ground

To successfully bridge the technical and business gaps in a start-up environment is like being the fulcrum between the two ends of a see-saw. Holding the center position allows you to value and consider both, with an eye to the higher ground where the solution lies.

- **Stay focused on the priority at hand.** People will challenge each other, personally and professionally, and you'll need to direct them away from squabbling and avoid getting pulled into long drawn out debates or heated disagreements that can drain you and your team's energy away from moving the company forward.

- **You are the decision maker.** At some point, while respecting everyone's input, you must don the leader's hat and make a decision, regardless of the pushback. As you are the head of the team, everyone will heavily rely on you.

- **Invest in your soft skills.** The way you communicate, the way you see problems, the way you manage people, the way you plan your day, the way you evaluate your short and long-term milestones are very different in a research lab—which is why you must invest in developing your people skills and self-awareness as they are the most important part of your job.

Next Up: SHOW ME THE MONEY

What funding strategies fit the stage of your start-up?

Chapter 8
SHOW ME THE MONEY, Part I

Raising money is the second hardest part of starting a start-up. The hardest part is making something people want: most start-ups that die, die because they didn't do that. But the second biggest cause of death is probably the difficulty of raising money. Fundraising is brutal.
– Paul Graham

Your first fundraising check marks the beginning of your journey as a viable company. And, as most founders would attest, it's one of the most critical—and challenging—aspects of a start-up. It enables you to do the work of development, without which you have no product and no company. In our early days, a series of small state grants and innovation awards, from $10,000 to $40,000, made it possible for us to develop our prototype. Later, a Phase I and a Phase II NIH grant put us on the road to commercialization. Through trial and error, over a period of three years, we raised over $1.4 million dollars of non-dilutive funding.

During that time, I went from lead chemist of a 4-person technical team to being the founder and CEO of Hydoglyde Coatings. There's no doubt about it: money makes you real. It gives you recognition in your space and the power to execute. Without it, all you have is an idea. It's the difference between having your canoe stuck on the muddy bank or being propelled forward by the current of the river.

So how will you begin your journey? Get that first check that puts you on the map and lets you grow your team, secure lab space, seek proper legal or accounting counsel, and secure the materials and equipment to validate your idea.

Like many early stage start-ups with a promising but unproven idea, especially those spun out of a university or ones that require extensive research and prototyping, initial seed or pre-seed funding often comes from grants. Other founders seek outside raises, especially if they have advanced beyond proof-of-concept and are ready to commercialize their technology. Others may pursue both, which is what we did. Whatever route you choose, fundraising is incredibly challenging and there is no one right way to do it. It just has to be done.

Let's take a more detailed look at your options and consider which strategies fit the stage of your start-up.

Types of Funding

The most common funding sources can be broken down into two types: non-dilutive and dilutive. Non-dilutive is any kind of funding that does not require you to give up ownership: grants and awards. Dilutive funding is any kind of funding which forces you to give up ownership: venture capital, angel investing. Of course you might also consider bank loans, money from friends and family, and fundraising platforms like GoFundMe. In this chapter, I'm going to focus on non-dilutive grants, particularly governmental Small Business Innovation Research (SBIR) and Small Business Technology Transfer (STTR) opportunities in which I was both a receiver and reviewer. In the next chapter, I'll discuss dilutive funding rounds, such as venture capital and angel investment. At the end of the book, you'll find a list to explore resources.

Non-dilutive funding: Grants

Early stage technical start-ups are often encouraged to go the grant funding route for a number of reasons. First, the golden word: non-dilutive. Unlike a traditional raise, you do not need to give a percentage of your company away when awarded.

Another key advantage of grants is that you can secure funding with just an idea (in addition to a well-thought out business plan strategy, of course). Grant opportunities, like SBIR and STTR, are designed to support technical founders with highly innovative and highly risky start-up ideas that have the potential to tackle a big problem, which in our case is the global epidemic of sexually transmitted diseases.

A grant (or series of grants) gives you the time—often years—to demonstrate the efficacy of your idea, to show the outside world (investors/partners) that your research has value and commercial potential. You and your technical team can test your hypothesis, develop a working prototype, or establish a proof-of-validation as you grow your company to a strong valuation so when the time comes, you're prepared to raise outside funding. From an idea on paper, it took us three years to develop a working prototype of a new condom coating technology in our goal to shake up a 70-plus year industry stagnation.

The SBIR/STTR grant

There are two main federal programs sponsored by the Small Business Agency (SBA) that support technology development and early commercialization: The Small Business Innovation Research (SBIR) and Small Business Technology Transfer (STTR). Both programs partner with and operate at state and federal levels and share the same goal: to support your innovative idea that has commercial potential. The difference between them is that the SBIR is for early stage start-up applicants taking the majority lead to carry out proposed aims and the STTR is focused on early stage start-up applicants that partner with a collaborator, usually from academia or research hospitals. If your science or tech start-up is in a state, such as MA, TX, OR, and CA, that offers follow-up opportunities to match awarded SBIR/STTR funding, you may be able to capture additional support.

Getting funding through these routes, however, is incredibly competitive. To begin, do your research and get a lay of the land so you know what's available. Prepare a comprehensive list, then decide on a strategy. You may need a consultant or outside advisor to help. Although we funded our start-up by securing state and federal grants, including a Phase I and Phase II SBIR from the NIH, I've submitted dozens of other proposals that were flat out rejected.

The competition gets stiffer and stiffer each year as funding agencies across all governmental branches (such as NIH, NSF, DOD, NASA) are receiving more applications than ever before. Although this can be daunting to some founders, you have to throw your hat into the ring for a winning shot. Just realize upfront that it's a long haul: preparation and patience is the name of the game.

If you are an early stage company, you may decide to submit a Phase I SBIR/STTR proposal in an attempt to secure $250,000. However to be competitive enough for a reviewer to be excited about your idea, a lot of preparation must go into a proposal. Depending on the agency, you will spend months preparing the 75-100 page application. It needs to cover everything from your idea and the problem it solves, your technology and how it is innovative, your background, who is on your team, and what you are going to do with the funding if awarded.

Our story

To get the SBIR grant—on our second try—we had to fulfill several requirements to demonstrate that our idea was feasible.

- Present our technical strategy—our detailed plan for how we were going to achieve proof of concept. What are the big two or three key experiments that would get us there?
- Identify our deliverables. Exactly what would we accomplish with the money? What are the specific actions and timelines? How would we know if we were successful in achieving them?
- Anticipate problems. What challenges will arise and how will we mitigate them? What is our Plan B?

- Showcase our strong team with the right talent and technical skillsets to do the job.
- Demonstrate the commercial potential of our idea in letters of support from reputable collaborators and potential end customers.

The fact that we had initial support from the Melinda and Bill Gates Foundation, along with some earlier grants, gave credence and support to our application. Does that mean you shouldn't apply right away? Not necessarily. Just realize that the credibility you build over time with smaller grants has an impact.

As a reviewer and consultant to dozens of companies seeking SBIR funding, a common mistake (especially first time SBIR/STTR candidates) is the lack of detail in the applications. Not only do you have to sell the reviewer on your idea but your job is to give them the confidence you have thought through every single detail of every single experiment you propose, all the way from how many replicates you plan to carry out in your assay, number of formulations tested, and each analysis of each prototype. The more details you provide, the stronger your application will be.

The waiting game

So you and your team have spent months preparing your SBIR/STTR application. You have re-read and revised it for the 20th time. You hit the send button and wait. It can be three, four, five, or even six months before learning the outcome of your proposal.

Once founders finally get the results of their hard work and labor, quite frequently (in fact most of the time) they are disappointed. Due to the highly competitive nature of these grants, the majority of first time SBIR/STTR pursuees do not get their applications funded the first time around.

What you will get are critiques and feedback from the review panel. Be forewarned that some of the comments and feedback may be harsh or highly critical. It's discouraging—and a shock for many founders—to have a reviewer tear apart their idea after spending months working on the application. But as we've all learned at the bench top in pursuit of our research, perseverance is the key to success. View the feedback as constructive criticism and learn from it. Use it as an opportunity to strengthen your business strategy and technical objectives. It's a great learning experience.

Often you can revise and resubmit your application, though you'll have to go through the waiting cycle again. But the good news is that typically these applications go back into the pool much stronger the second, third, or even fourth time around as founders gain a better understanding of what reviewers are looking for and how they are evaluating their start-up, technology and application.

Due to the frantic and stressful event of fundraising, many first-time founders tend to reuse grant applications for different grant opportunities. While this is encouraged, it is important to make sure all grant applications are aligned to the exact call of the proposal. There are too many times, both as a grant consultant and reviewer, that I have seen companies trying to force their technology or solution into an area that may not be applicable at all.

Next Up: SHOW ME THE MONEY, Part II

When is it time to do an outside raise? What are the basic preparations?

Chapter 9
SHOW ME THE MONEY, Part II

If the business were a play, Act One is: Woohoo, bright and bushy-tailed.
We're going to make something great!
Act Two is: We're six months behind on back-end development.
We're trying to raise venture capital.
- Caterina Fake, Founder of Flickr and Partner, Yes VC

We've all heard the sensational stories of venture capital. Two guys with an idea get a million dollars to bring it to market and in a few years they get acquired by big tech for 25x the investment. Everybody goes off into the sunset, their dreams fulfilled. Except that it hardly ever happens this way. What does happen is that even with outside capital, 90% of start-ups fail. If you are a female or minority founder (such as myself), statistics show that your odds are unfortunately even lower. It's sobering—and yet, not a reason to keep you from pursuing a raise. There is a lot of money out there and if you find an investor to back you up and believe in you, the money can come in quickly and give you the opportunity to prove your start-up. So while looking clear-eyed at the difficulty—the competition, the risks, the odds—you have to keep forging ahead.

The fundraising mindset

The key is to get into the right mindset from the beginning—which is to understand where the heart of the challenge lies. In a nutshell: Successful fundraising depends on your ability to tell your story (assuming you've got a good one with evidence-based data and support to back it up.) What you have to accomplish is to transfer your belief and commitment about your start-up to another person or persons so they can see your vision and share your enthusiasm.

Think about it.

You've been dreaming about—and living your idea for a long time. After extensive research and experimentation, you have proven to yourself that it's not only feasible, your venture deserves support. You have people on your team who also believe in it. You have committed yourself and now the biggest challenge you face is to transfer your vision to someone else, to recreate the excitement, enough so that someone, an actual investor, will put down money on the table for you to pursue it. The rocket is on the launch pad and you need fuel to shoot it off into space.

So, at ground control: Are you ready for a capital raise? Let's look at the pre-requisites.

Pursuing an outside raise

Start-ups can raise at any time, even at the beginning of their formation, even without an idea. Of course this comes with a major downside since your company is valued at $0 and investors will take a significant chunk of your equity (sometimes even more than 50%) because you are a super high risk investment. Usually these funding rounds are called pre-seed or seed because the goal is to just get your idea off the ground. For some start-ups, this may be the best way to go. However for many others, especially in the research and tech space, it may be wise to hold off on venture capital (VC) and consider non-dilutive routes until they have the data and market support to back up their idea or innovation and thus build value in their start-up.

Sometimes start-ups decide to go directly to an outside raise to avoid the months of waiting involved in a grant process. For an early stage founder desperate to find money, this route can be tempting. But keep in mind that finding the right match is also a time-consuming process, one which can re-direct your valuable resources away from your core business activities. And if you do succeed, once you accept the money, investors *become a part of your company.* They may want a board seat or direct you to hire someone who is not a good fit your company. This is a big deal, so make sure that it's the right investor who, besides funding, can also contribute valuable business knowledge and experience.

Partnering with an investor is a marriage and sometimes one that is harder to get a divorce from than your spouse.

Most companies seek an outside raise when they are ready to establish a commercial market. Typically, you've gone beyond concept development and you have a working prototype. In other words, from the investor point of view, you have "de-risked" your technology. When you reach that stage, investor capital can make it possible for you to scale rapidly. Of course, you will pay for this kind of support by giving away a portion of your company and perhaps some of its control.

Rule of thumb: Don't seek outside capital until you have a strong valuation. Without a product or demonstrated market, a seed or pre-seed start-up is usually too high a risk for most VCs. Your company will be enticing to an investor when they see that you've already received grant funding or other market support that demonstrates your value. It can definitely move you to the front of the line.

Our story

During my very early days maneuvering the start-up world as a first-time founder, while I was tirelessly writing grants in hopes of securing non-dilutive funding, I was seeking outside capital at the same time. I threw my darts and knocked on dozens of VC and angel's doors and pitched at tons of meetings throughout the Boston area. Our prototype was scrappy and not quite there yet. We just completed our cap table and formation documents. And, we were getting desperate for additional funding to continue our developments. Meeting after meeting, I kept getting the same feedback:

"Come back next quarter."

"I love what you are doing but we do not have wiggle room right now."

"Keep us updated with your developments."

Turns out hindsight was 20/20, and this is actually how investors (kindly) say no. This was a lesson I had to learn quickly to realize we had a better chance to fund our company through grants at that stage. Still, I'm glad I started these conversations early. To build confidence with investors takes time and relationship building. What they want to see is persistence and, most importantly, progress and lots of it. As we've progressed through our developments and built value in the company, I've kept in contact with these connections by updating them regularly, making it easy to tap into this network later and pull the trigger whenever I am ready to raise.

The other benefit: I got much better at pitching. Meeting after meeting, I had to get in there and give my pitch over and over again. After the 20th time, I started noticing the same patterns of questions from both VCs and angels—and progressively, my answers became more sophisticated. With each meeting, I also received valuable feedback that helped me see my start-up in a new light that I hadn't considered before. Investors and angels were often generous with their time, brainstorming ideas with me and showing me how to structure my start-up.

Here's some of the game-changing advice we were given:

- Grow your team beyond its technical expertise to position the company for a strong investment and potential acquisition down the road. (At that time we were a team of only technical founders.)
- Consider expanding your technology for adoption into other products to generate other revenue sources. (We were focused on a single product and had not considered other uses for the coating technology we had developed.)
- Restructure the cap table appropriately so that you don't get diluted as a lead founder when you're ready to raise outside funding. Make sure each member of the cap table has a proper split appropriate to their future contribution and work to be done. Investors can learn a lot about a company

just by glancing at a cap table. (Later we revised our cap table to be more equitable.)

- Revise your pitch to focus on the business proposition and less on the technical slides, in order to attract investors. If they are interested in the heavy technical aspects, they will follow-up with you afterwards. (Further down the road, we now have a business strategy that reflects the marketplace opportunities.)

- Learn about the supply chain and how changing global economic and market conditions affect the price of goods, your market opportunities, and product introduction.

Seek alignment

Whatever outside funding route you decide to pursue make sure it's aligned to your own mission, problem statement, vision, and goals for your start-up. Create a target list and before meeting with VC firms or angels, find out what types of companies they have previously funded or are interested in supporting. Many firms list their portfolio on their website and you can also search sites like CrunchBase and AngelList. Make sure the firm also invests in the stage of funding you're looking for since some may specialize in pre-seed or seed investments whereas other firms focus on Series B or C for more established start-ups.

Develop a pitch that shows that you have a truly unique value proposition. Have a business plan, competitive analysis figures, pro forma projections, and backup slides ready to go. And most importantly, be prepared for the questions they will ask. They will try to find your weaknesses and be less interested in your strengths. Your weaknesses, if significant, will dismiss you immediately. Questions might include:

What is it about your approach, your technology and your team that makes your company able to achieve and execute on this opportunity?

What's your current value proposition?

Who are your customers and how are they different than your end users?

What is your plan to grow?

Why are you the right person to lead this company?

Handling rejection

Investors sit through hundreds of pitches each month and conduct due-diligence with dozens of start-ups every year. These are very expert people that will offer feedback and constructive criticism to challenge you to think of your start-up in a different light so that when you go back into the tank, you will go in much stronger, much wiser, and much more prepared.

However, there may be times that the feedback (from an investor or grant reviewer) totally bombs your idea! In fact, several people laughed outright in my face when they heard ours, thinking it was completely absurd. As the founder, this is not the kind of feedback to take seriously. Stand your ground—not everyone will get you, your company, or your vision. This is a difficult thing to do with rejection after rejection after rejection. Keep in mind that it's a long road and along the way, it's natural to doubt your idea or even yourself. If this happens, do two things.

First: turn to the people, colleagues, team members and community who support you. They understand the struggles you're going through and they have your back. Also, keep in mind that the #1 person you need support from on this journey is yourself.

Second: keep going. Persistence is the name of the game. There are thousands of financial opportunities out there—and you only need one to start! Once you capture that first dollar, dilutive or non-dilutive, you will be rewarded with one of the highest highs in the start-up world.

LESSONS LEARNED: Showcase a solid and convincing business proposition

The start-up vision in your mind may be a sure thing. However to an investor that is highly skeptical, your job is to make sure you present a strong proposition and opportunity for them to hop onboard. You have to tell a persuasive and bullet-proof story to penetrate an investor's natural defenses against it.

- **The basics.** Whatever type of funding you decide to focus on, dilutive or non-dilutive, to give your start-up the best chance, you need to:
  1) Convince you have the best solution to solve a problem.
  2) Demonstrate that your idea has a competitive edge.
  3) Showcase a strong team.
  4) Plan out your short- and long-term goals with well-thought out milestones.

- **You have options.** In deciding what kind of funding to pursue, realize that you have options: venture capital, angel investors, and any number of local, state, or federal grants. Come up with a strategy so you can seek a balance: too many darts thrown at too many targets can lead to burnout.

- **The first funding is the hardest.** Getting the first grant or securing the first investor is the hardest. It is a hunt to find that one organization or the one investor that will take a chance on you, your idea, and the start-up.

- **Listen to the (right) critiques.** Whether it's negative feedback from reviewers of your 100-page grant or a rejection from a potential investor, it can be hard to take. It is important to learn not to take these negative critiques personally! Keep in mind that these are very expert people whose evaluation is meant to help you. Use it to challenge yourself to see your start-up with a new perspective.

- **Do your homework.** Whatever funding opportunity you decide to go for, make sure it is aligned to your own mission, problem statement, vision, and goals for your start-up.

- **Get your documents in order.** Don't underestimate the preparation of documents (for grants or a capital raise) that tell your story and demonstrate your competitive edge, including: pitch deck, business plan, a one-page investment summary, or term sheet. You would always rather be over-prepared than under-prepared.

Next Up: ONE THOUSAND TRIES

Building your idea into a real company can be a long and discouraging road—how do you keep going?

Chapter 10
ONE THOUSAND TRIES

No one can build you the bridge on which you, and only you, must cross . . . There is one path in the world that none can walk but you. Where does it lead? Don't ask, walk!
— Friedrich Nietzsche, Philosopher

It took me and my founding team over three years and (literally) 1,000 tries to get to a prototype formulation that was even remotely close to commercialization. Toward the end of that long and discouraging road, there were many voices that expressed doubt. Gradually, the number of failures and dead-ends accumulated. The team and I became unsettled and so, so weary. As we were rounding the last bend, I faltered. I felt like giving up. Our goal was to disrupt a stagnant industry whose basic product hadn't changed in 70 years. Many times I thought it wasn't going to happen. There were (many) nights after rough days when I looked for comfort from a bottle of wine or just ended up burying my face into a box of tissues to let the waterfall go. Hell, we'd put in the time—over 10,000 hours—and done our best. Was I just being stubborn for continuing? Maybe it was time to throw in the towel.

And yet, there was something, an urge beyond all the evidence, beyond the logical argument of our trusted advisors, family, and friends that I fully agreed with. Something kept pushing us, a gut feeling we had. So the team met and collectively decided to pursue one more prototyping round. The next Saturday afternoon, mentally prepared for another failed experiment, without enthusiasm, I mixed a solution and created the longed-for result: a slippery latex surface. Fast-forward three years: we moved toward a manufacturing round, passed preliminary biocompatibility testing, and now we're in human trials.

In closing, I want to talk about the one thing necessary, the elusive urge, the inner knowing that can seem completely off—or independent from what the logical scientist and savvy business mind tells us. There are many voices—external and internal—that can drown out this urge, this knowing. The dire warnings of others who've walked the path and know the journey, whose experience extends far beyond yours, can be especially overwhelming.

To find the inner voice, you have to get quiet. You have to get calm. Step back from your fears, allow the ripples of other voices to fade and throw another stone out onto the pond. Watch and feel what happens in that circle of new undulations. What is true? What do you want to do? What is right, *for you*?

No matter where you are on your start-up journey—at the germ of an idea, a working prototype, your first seed funding, or what seems like a dead end—to move forward, there's one thing necessary you must always return to. Yourself.

Developing an entrepreneurial mindset

Failure is bad. It's been drilled into our heads. We must avoid it at all times. It's something to be ashamed of, to be hidden from others. The truth is that our mistakes and missteps, what we might call our failures, are part of being an entrepreneur.

I believe that the only way to really fail is not to try.

Going from Point A to Point B is not a straight line. There are lots of twists and pivots along the way. When you encounter a "dead end," it doesn't mean that you should stop your investigation or give up on your start-up idea. It means you need to take a step back and reassess the situation. Why did X, Y or Z yield this outcome instead of the one I expected? Then try again. Remember that history is filled with major scientific and technical advancements that have come from so-called "failed experiments."

On a practical day-to-day level, however, we do have to deal with the discouragement, the doubt, and the feeling of failure. How do you get through the dark days and stay motivated? First, recognize that your success does not rest on your idea alone. It depends on something more elusive, something invisible whose power you might not be aware of, something that can't be charted or measured by data: Your mindset. The lens you habitually look through every day will move you forward or stall you out. It's what makes you an entrepreneur (or not).

Let's look at the basic strategies that support this mindset.

> **Face your fears.** Look specifically at your fears and put them on the table. Examine and question them, but avoid spinning out stories of disaster. Then revisit your long-term goal. Revalidate yourself and go back to your why. Why are you doing this? How important is it to you?

Recognize your negative thinking. Negativity can become a dangerous spiral. It will undermine your ability to be creative and to lead your team. One way to shift your thinking is to look at a negative pain point and ask "How can I actively correct or respond to this problem?" Maybe it's a discussion you've been avoiding but need to have with your team or support network.

Don't ignore self-care. Long hours and erratic demands will negatively impact your mental and emotional outlook. Self-care is a life skill that goes a long way to helping you stay healthy and maintain a positive mood. Eating well, sleeping enough, exercise and taking regular breaks are things you can control. Even a 10-minute walk can make a difference in your attitude, boost your mood and reduce negative thoughts.

Never lose your curiosity. Keep feeding your curiosity and trying new ideas. The danger is stopping and getting stuck in parking mode. To move forward, you have to turn the key in the ignition and give it fuel. Stop rehearsing the what-ifs and move on to the next step. You might not be able to see around the corner, but you *can* take the next step.

Go for the ultimate positive attitude. I continue to be inspired and energized by self-motivated entrepreneurs, encouraging mentors, and savvy

> investors who demonstrate what I call the Ultimate Positive Attitude. These are people who have who have built and sold multiple successful businesses, raised billions of dollars, and engaged in groundbreaking research. These same people have tackled some of the world's biggest issues from cancer research to improving medical devices and energy sustainability.

This ability is actually a decision you can make every day, a decision to regard every situation as leading you where you need to go, whatever the challenge, the mistake, the rejection, the dead-end. In this view you can't fail because every step you're taking is toward your ultimate destination. That's part of the adventure. A mistake you make today may lay the ground work for opportunities you cannot even imagine tomorrow.

Crossing the invisible bridge

An iconic scene for any adventurer or explorer of the unknown—and one of my favorites—is the literal leap of faith Indiana Jones makes in *The Last Crusade* in his search for the Holy Grail. He emerges from the opening in the rock face and looks down at the black bottomless abyss, then looks at his small book of instructions. He looks again at the abyss and says, it's impossible, nobody can jump this. Another voice calls out: You must hurry, come quickly. Fear and trembling take hold of him. The image of his father appears in his mind telling him to believe, to have faith. He closes his eyes and rests his hand on his heart. Then the all-important moment: complete stillness. When he opens his eyes, something has changed. Beyond logic or sense of any kind, he lifts his leg and extends it forward as if to step out on the air itself. His foot makes contact with the bridge that only becomes visible after he steps out on it.

I hope this book has given you a bridge to support your idea and take you forward on your start-up journey: to give you a better understanding of the obstacles, the challenges, and the opportunities ahead.

Success on the journey,

Stacy Chin, Ph.D.

NOTES

Chapter 4
Steven Blank, "Why scientists make bad entrepreneurs and how to change that."
https://qz.com/502143/why-scientists-make-bad-entrepreneurs-and-how-to-change-that/

"The Global Medical Devices Report"
https://www.globenewswire.com/news-release/2020/10/27/2114984/0/en/Global-Medical-Device-Market-2020-Size-To-Increase-Due-To-Rising-Infectious-And-Chronic-Disease-Cases-As-Per-The-Business-Research-Company-s-Medical-Devices-Global-Market-Opportuni.html

Chapter 5
Study by the Massachusetts Biotechnology Foundation, *Scientific American*

https://blogs.scientificamerican.com/observations/soft-skills-in-the-life-sciences/

Seth Godin, "Let's stop calling them soft skills."
https://itsyourturnblog.com/lets-stop-calling-them-soft-skills-9cc27ec09ecb

Chapter 8
Small Business Innovation Research (SBIR) and Small Business Technology Transfer (STTR) https://www.sbir.gov/about

Chapter 10
"The Last Crusade," Bridge scene.
https://www.youtube.com/watch?v=sBBbq2g7yf8

RESOURCES

Websites

Scientific American
https://blogs.scientificamerican.com/observations/soft-skills-in-the-life-sciences/
Seth Godin's Blog https://itsyourturnblog.com/lets-stop-calling-them-soft-skills-9cc27ec09ecb

On Fundraising
Crunchbase https://www.crunchbase.com/1

Small Business Administration https://www.sba.gov/
Small Business Innovation Research https://sbir.nih.gov/
America's Seed Fund https://seedfund.nsf.gov/
Defense Innovation Portal
https://www.dodsbirsttr.mil/submissions/login
Angel Investors https://angel.co/

Books

Dale Carnegie, *How to Win Friends and Influence People (Paperback)*

Stephen R. Covey, *The 7 Habits of Highly Effective People: Powerful Lessons in Personal Change (Paperback)*

Daniel Kahneman, *Thinking, Fast and Slow (Hardcover)*

Roger Fisher and William L. Ury, *Getting to Yes: Negotiating Agreement Without Giving In (Paperback)*

Keith Ferrazzi, *Never Eat Alone: And Other Secrets to Success, One Relationship at a Time (Hardcover)*

James Clear, *Atomic Habits: An Easy & Proven Way to Build Good Habits & Break Bad Ones (Kindle Edition)*

Jim C. Collins, *Good to Great: Why Some Companies Make the Leap... and Others Don't (Hardcover)*

Mark Goulston, *Just Listen: Discover the Secret to Getting Through to Absolutely Anyone (Hardcover)*

Made in the USA
Monee, IL
04 November 2026

67385885R10056